Number Theoretic Algorithms

N.B. Singh

DEDICATION

To Nature,

I dedicate this book to you, the source of all life. You are my inspiration, my teacher, and my friend.

Thank you for teaching me about the beauty of the world around me. Thank you for showing me the power of the natural world. Thank you for giving me a sense of peace and tranquillity.

I promise to do my part to protect you and your many wonders. I will teach my children about the importance of conservation and sustainability. I will work to make the world a better place for all living things.

Thank you for everything, Nature.

With love,

N.B Singh

Contents

PREFACE

Welcome to "Number Theoretic Algorithms," a comprehensive exploration of the fascinating world of algorithms in number theory. This book aims to provide readers with a deep understanding of the mathematical principles and computational techniques that underpin the study of integers.

Motivation

Number theory, the study of integers and their properties, has been a cornerstone of mathematics for centuries. Its elegance and simplicity hide a world of complexity and depth, making it an ideal playground for algorithmic exploration. The motivation behind this book is to unravel the intricate algorithms that power the solutions to problems in number theory.

Scope and Organization

The book is organized into several chapters, each focusing on a specific aspect of number theory and the algorithms associated with it. We begin with fundamental concepts such as divisibility, prime numbers, and modular arithmetic, gradually progressing to more advanced topics like public-key cryptography, elliptic curves, and algebraic number theory.

Target Audience

This book is designed for students, researchers, and practitioners in mathematics and computer science who have an interest in the beauty and utility of number theoretic algorithms. While a basic understanding of number theory is helpful, the book is structured to be accessible to a wide audience, including those with a strong algorithmic background.

Features of the Book

- Clear explanations of key number theoretic concepts and algorithms.

- Numerous examples to reinforce understanding.

- Insights into the practical applications of number theory in various fields.

- Discussions on the historical context and development of algorithms.

Chapter 1

Introduction

1.1 Background

Number theory is a branch of mathematics that explores the properties and relationships of integers. It has deep historical roots and has been a fundamental area of study for centuries. One of the central concepts in number theory is the notion of primes.

A prime number is an integer greater than 1 that has no positive divisors other than 1 and itself. For example, 2, 3, 5, and 7 are prime numbers. The Fundamental Theorem of Arithmetic states that every positive integer greater than 1 can be uniquely expressed as a product of prime numbers.

$$n = p_1^{e_1} \cdot p_2^{e_2} \cdot \ldots \cdot p_k^{e_k}$$

Here, $p_1, p_2, \ldots, p_k$ are prime numbers, and $e_1, e_2, \ldots, e_k$ are their respective exponents. This unique factorization property forms the basis for many number theoretic algorithms.

Another important concept is modular arithmetic. In modular arithmetic, we consider the remainder when dividing one integer by another. The notation $a \equiv b \pmod{m}$ indicates that a and b have the same remainder when divided by m.

$$7 \equiv 1 \quad (\text{mod } 3)$$

Modular arithmetic finds applications in various cryptographic algorithms, where the periodicity of remainders is leveraged for secure communication.

The Euclidean Algorithm, a cornerstone of number theory, computes the greatest common divisor (GCD) of two integers. The GCD is the largest positive integer that divides both numbers.

$$\text{GCD}(a, b) = \text{GCD}(b, a \bmod b)$$

Let's consider an example: finding the GCD of 48 and 18 using the Euclidean Algorithm.

$$\begin{aligned}
\text{GCD}(48, 18) &= \text{GCD}(18, 48 \bmod 18) \\
&= \text{GCD}(18, 12) \\
&= \text{GCD}(12, 18 \bmod 12) \\
&= \text{GCD}(12, 6) \\
&= \text{GCD}(6, 12 \bmod 6) \\
&= \text{GCD}(6, 0) \\
&= 6
\end{aligned}$$

This algorithm is essential for various applications, including reducing fractions to their simplest form.

The study of number theoretic algorithms is not only theoretical but has practical applications in computer science, cryptography, and information security. In the following chapters, we will delve into more advanced algorithms and explore their wide-ranging applications.

1.2 Motivation

Motivation serves as the driving force behind the exploration of number theoretic algorithms. One primary motivation lies in the field of cryptography, where the security of communication relies on the complexity of certain mathematical problems. Number theory, with its intricate algorithms, plays a pivotal role in creating secure cryptographic systems.

One such algorithm that motivates the study of number theory is the RSA algorithm. RSA relies on the difficulty of factoring the product of two large prime numbers. Let p and q be distinct prime numbers, and let $n = pq$. The security of RSA hinges on the difficulty of factoring n back into p and q.

$$n = pq$$

To motivate further, consider the public-key cryptography system. Each user has a public key for encryption and a private key for decryption. The security of the system relies on the fact that, while it is easy to multiply two large primes to obtain n, factoring n into its prime components is computationally infeasible for large enough primes.

The study of modular arithmetic also plays a vital role in the motivation behind number theoretic algorithms. Modular arithmetic is essential for designing secure hash functions and pseudorandom number generators. The periodicity and unpredictability of modular operations contribute to the robustness of these cryptographic systems.

$$a \equiv b \pmod{m}$$

As an example, consider the use of modular arithmetic in the generation of pseudorandom numbers. By choosing a suitable modulus and generator, one can create sequences of seemingly random numbers that are deterministic and repeat after a certain period. This property is valuable in various applications, including simulations and cryptographic protocols.

Motivation for number theoretic algorithms extends beyond cryptography into the realms of computer science and information technology. Efficient algorithms for number theory problems have implications for optimizing computer programs, enhancing data security, and advancing various fields.

Chapter 2

Fundamental Concepts

2.1 Prime Numbers

Prime numbers are the building blocks of number theory and play a central role in many number theoretic algorithms. A prime number is defined as a positive integer greater than 1 that has no positive divisors other than 1 and itself. The significance of prime numbers lies in their unique properties and their crucial role in the Fundamental Theorem of Arithmetic.

The Fundamental Theorem of Arithmetic states that every positive integer greater than 1 can be expressed as a unique product of prime numbers. This decomposition is often referred to as the prime factorization.

$$n = p_1^{e_1} \cdot p_2^{e_2} \cdot \ldots \cdot p_k^{e_k}$$

Here, $p_1, p_2, \ldots, p_k$ are prime numbers, and $e_1, e_2, \ldots, e_k$ are their respective exponents. Let's consider an example to illustrate this theorem.

Consider the number 120. Its prime factorization is given by:

$$120 = 2^3 \cdot 3 \cdot 5$$

This unique factorization property has profound implications for number theory and forms the basis for various algorithms.

One such algorithm is the Sieve of Eratosthenes, a method for finding all prime numbers up to a given limit. The algorithm works by iteratively marking the multiples of each prime, gradually eliminating composite numbers and leaving only the primes.

For instance, applying the Sieve of Eratosthenes to find primes up to 30:

$$2, 3, 5, 7, 11, 13, 17, 19, 23, 29$$

Prime numbers are not only important in theory but also find practical applications, particularly in cryptography. The security of many cryptographic systems, such as RSA, relies on the difficulty of factoring large composite numbers into their prime components.

Understanding the distribution of prime numbers is another fascinating aspect of number theory. The Prime Number Theorem, formulated by Jacques Hadamard and Charles Jean de la Vallée-Poussin independently, describes the asymptotic distribution of prime numbers. It states that the density of primes around a large number n is approximately $\frac{1}{\ln(n)}$.

2.2   Divisibility

Divisibility is a fundamental concept in number theory that explores the relationships between integers based on their ability to divide evenly. Understanding divisibility is crucial for the development of number theoretic algorithms and plays a central role in the study of prime numbers, modular arithmetic, and many other areas.

A basic definition of divisibility is that an integer a is divisible by another integer b if there exists an integer c such that $a = b \cdot c$. We denote this relationship as $b \mid a$.

$$b \mid a \iff \exists c : a = b \cdot c$$

As an example, 15 is divisible by 3 because $15 = 3 \cdot 5$. Mathematically, we express this as $3 \mid 15$.

Divisibility rules provide efficient ways to determine whether one number is divisible by another without explicitly performing the division. For example, a number is divisible by 2 if its last digit is even, and it is divisible by 3 if the sum of its digits is divisible by 3.

Consider the number 468. The sum of its digits is $4 + 6 + 8 = 18$, and since 18 is divisible by 3, we can conclude that 468 is also divisible by 3.

The concept of greatest common divisor (GCD) is closely related to divisibility. The GCD of two integers a and b, denoted as $GCD(a, b)$, is the largest positive integer that divides both a and b. Euclid's Algorithm provides an efficient way to compute the GCD.

$$GCD(a, b) = GCD(b, a \bmod b)$$

Let's find the GCD of 48 and 18 using Euclid's Algorithm:

$$
\begin{aligned}
GCD(48, 18) &= GCD(18, 48 \bmod 18) \\
&= GCD(18, 12) \\
&= GCD(12, 18 \bmod 12) \\
&= GCD(12, 6) \\
&= GCD(6, 12 \bmod 6) \\
&= GCD(6, 0) \\
&= 6
\end{aligned}
$$

Understanding divisibility properties is essential in modular arithmetic. In modular arithmetic, we consider the remainder when dividing one integer by another. If $a \equiv b \pmod{m}$, it means that a and b have the same remainder when divided by m.

$$a \equiv b \pmod{m}$$

For instance, if we consider $17 \equiv 5 \pmod{6}$, it means that both 17 and 5 leave the same remainder when divided by 6.

Divisibility concepts extend beyond basic arithmetic into the realm of algebraic structures, such as rings and fields. The study of these structures deepens our understanding of number theory and contributes to the development of advanced algorithms.

Chapter 3

Basic Algorithms

3.1 Euclidean Algorithm

The Euclidean Algorithm is a fundamental algorithm in number theory for finding the greatest common divisor (GCD) of two integers. Named after the ancient Greek mathematician Euclid, this algorithm provides an efficient way to compute the GCD and has applications in various mathematical and computational contexts.

The algorithm is based on the observation that the GCD of two numbers remains the same if we replace the larger number with its remainder when divided by the smaller number. Mathematically, it can be expressed as:

$$\text{GCD}(a, b) = \text{GCD}(b, a \bmod b)$$

This recursive property allows us to iteratively reduce the problem until we reach a base case. Let's illustrate the Euclidean Algorithm with an example: finding the GCD of 48 and 18.

$$\text{GCD}(48, 18) = \text{GCD}(18, 48 \bmod 18)$$
$$= \text{GCD}(18, 12)$$
$$= \text{GCD}(12, 18 \bmod 12)$$
$$= \text{GCD}(12, 6)$$
$$= \text{GCD}(6, 12 \bmod 6)$$
$$= \text{GCD}(6, 0)$$
$$= 6$$

The algorithm terminates when one of the numbers becomes zero, and the GCD is then the non-zero remaining number.

The efficiency of the Euclidean Algorithm stems from its ability to reduce the size of the input in each step. This makes it particularly useful for large numbers and forms the basis for more advanced algorithms in number theory.

Euclidean Algorithm has applications beyond finding the GCD. It is crucial in modular arithmetic, where it is used to find the modular inverse. The modular inverse of an integer a modulo m is an integer b such that $ab \equiv 1 \pmod{m}$.

Another application is in simplifying fractions. Given two integers a and b, the fraction $\frac{a}{b}$ can be simplified by dividing both the numerator and denominator by their GCD.

The extended version of the Euclidean Algorithm can also be used to express the GCD as a linear combination of the input numbers. For a, b with GCD d, there exist integers x and y such that $ax + by = d$.

$$ax + by = \text{GCD}(a, b)$$

Understanding the Euclidean Algorithm provides a solid foundation for more advanced algorithms in number theory, such as the Extended Euclidean Algorithm and applications in cryptography.

3.2 Sieve of Eratosthenes

The Sieve of Eratosthenes is a classic algorithm for finding all prime numbers up to a given limit. Named after the ancient Greek mathematician Eratosthenes, this algorithm efficiently eliminates non-prime numbers, leaving only the primes. It is a fundamental tool in number theory and has applications in various mathematical and computational contexts.

The algorithm works by iteratively marking the multiples of each prime, starting from the smallest prime, 2. The unmarked numbers remaining after this process are prime. Let's delve into the steps of the Sieve of Eratosthenes using an example to find primes up to 30.

Initially, all numbers are marked as potential primes.

23 45 67 89 1011 1213 1415 1617 1819 2021 2223 2425 2627 2829 30

Start with the first unmarked number, 2. Mark all multiples of 2.

23 45 67 89 1011 1213 1415 1617 1819 2021 2223 2425 2627 2829 30

Move to the next unmarked number, 3. Mark all multiples of 3.

23 45 67 89 1011 1213 1415 1617 1819 2021 2223 2425 2627 2829 30

Continue this process until you reach the square root of the limit. The remaining unmarked numbers are primes.

23 45 67 89 1011 1213 1415 1617 1819 2021 2223 2425 2627 2829 30

The primes are 2, 3, 5, 7, 11, 13, 17, 19, 23, 29.

The Sieve of Eratosthenes is highly efficient for finding primes, and its time complexity is close to linear. It forms the basis for various algorithms and is an essential topic in the study of number theoretic algorithms.

Chapter 4

Advanced Algorithms

4.1 Modular Arithmetic

Modular arithmetic is a foundational concept in number theory that deals with the remainders of integer division. It plays a crucial role in advanced number theoretic algorithms, cryptography, and various computational applications.

At its core, modular arithmetic is concerned with numbers that have the same remainder when divided by a fixed positive integer m. The notation $a \equiv b$ (mod m) denotes that a and b have the same remainder when divided by m.

$$a \equiv b \pmod{m}$$

One fundamental property of modular arithmetic is that addition, subtraction, and multiplication are well-defined operations. For any integers a, b, c, we have:

$$(a + b) \mod m \equiv ((a \mod m) + (b \mod m)) \mod m$$

$$(a - b) \mod m \equiv ((a \mod m) - (b \mod m)) \mod m$$

$$(a \cdot b) \mod m \equiv ((a \mod m) \cdot (b \mod m)) \mod m$$

For example, consider $17 + 25 \equiv 12 \pmod{10}$. Here, both $17 \mod 10$ and $25 \mod 10$ are equal to 7, and their sum is 14, which leaves a remainder of 4 when divided by 10.

Modular exponentiation is a critical operation in number theory and cryptography. It involves computing $a^b \mod m$, where a, b, and m are integers. This operation is efficiently performed using algorithms like exponentiation by squaring.

Let's illustrate with an example: $3^{10} \mod 7$. Using exponentiation by squaring:

$$
\begin{aligned}
3^{10} &= (3^5)^2 \\
&= (243 \mod 7)^2 \\
&= 6^2 \\
&= 36 \mod 7 \\
&= 1
\end{aligned}
$$

Thus, $3^{10} \equiv 1 \pmod{7}$.

Modular inverses are another crucial aspect of modular arithmetic. The modular inverse of a modulo m, denoted as a^{-1}, is an integer b such that $ab \equiv 1 \pmod{m}$. Not all integers have modular inverses, but if a and m are coprime, an inverse exists.

As an example, let's find the modular inverse of 3 modulo 7:

$$
\begin{aligned}
3 \cdot 5 &\equiv 1 \pmod{7} \\
15 &\equiv 1 \pmod{7}
\end{aligned}
$$

So, the modular inverse of 3 modulo 7 is 5.

Modular arithmetic has numerous applications in cryptography, where it forms the basis for algorithms like the RSA algorithm. Understanding its properties is crucial for designing secure and efficient cryptographic systems.

4.2   RSA Algorithm

The RSA algorithm, named after its inventors Ron Rivest, Adi Shamir, and Leonard Adleman, is a cornerstone in modern cryptography. It is a public-key cryptosystem that relies on the mathematical properties of number theory, particularly modular arithmetic.

The key idea behind RSA is the use of the difficulty of factoring large composite numbers into their prime components. The algorithm involves the generation of two large prime numbers, p and q, and the computation of their product $n = pq$. The security of RSA is based on the practical impossibility of factoring n back into p and q when n is sufficiently large.

$$n = pq$$

To establish the public and private keys, RSA uses the totient function $\phi(n)$, where $\phi(n) = (p-1)(q-1)$. The public key consists of n and an exponent e, where e is typically chosen as a small prime number. The private key includes n and a private exponent d, which is the modular multiplicative inverse of e modulo $\phi(n)$.

The encryption process involves raising a message M to the power of e modulo n, resulting in the ciphertext C:

$$C \equiv M^e \pmod{n}$$

As an example, let's encrypt a message $M = 42$ with a public key ($n = 187, e = 7$):

$$C \equiv 42^7 \pmod{187}$$

The decryption process involves raising the ciphertext C to the power of d modulo n, recovering the original message M:

$$M \equiv C^d \pmod{n}$$

Using the private key corresponding to the chosen public key, we can decrypt the ciphertext:

$$M \equiv C^d \pmod{187}$$

The RSA algorithm ensures that only the possessor of the private key can decrypt the message. Its security relies on the difficulty of factoring n into p and q, even with the knowledge of the public key.

RSA finds applications in secure communication, digital signatures, and the establishment of secure channels in various online protocols. Its robustness against classical and quantum attacks makes it a widely adopted cryptographic algorithm.

Chapter 5

Applications

5.1 Cryptography

Cryptography, the art and science of secure communication, heavily relies on number theoretic algorithms for providing privacy and integrity in data transmission. In this section, we explore the fundamental principles of cryptography and how number theory plays a crucial role.

One of the key concepts in cryptography is the use of modular arithmetic, particularly in the context of public-key cryptography. Public-key cryptography employs a pair of keys: a public key for encryption and a private key for decryption.

The RSA algorithm, a prominent example, utilizes modular arithmetic and prime factorization. Let n be the product of two large prime numbers p and q, and $\phi(n) = (p-1)(q-1)$ be the totient function. The public key is (n, e), and the private key is (n, d), where d is the modular multiplicative inverse of e modulo $\phi(n)$.

Encryption: $C \equiv M^e \pmod{n}$, where C is the ciphertext, M is the message.

Decryption: $M \equiv C^d \pmod{n}$.

For instance, consider encrypting a message $M = 88$ using a public key $(n = 187, e = 7)$:

$$C \equiv 88^7 \pmod{187}$$

The security of RSA lies in the difficulty of factoring n back into p and q, even with knowledge of the public key.

Another widely used cryptographic algorithm is the Diffie-Hellman key exchange, which establishes a shared secret key between two parties over an insecure communication channel.

Let g be a generator and p be a prime number. Party A chooses a secret key a and sends $A \equiv g^a \pmod{p}$ to Party B. Similarly, Party B chooses a secret key b and sends $B \equiv g^b \pmod{p}$ to Party A. The shared secret key is then $K \equiv A^b \pmod{p} \equiv B^a \pmod{p}$.

Cryptography extends beyond public-key systems. Symmetric-key cryptography uses a single key for both encryption and decryption. Advanced Encryption Standard (AES) is a widely adopted symmetric-key algorithm.

Consider a block of plaintext P and a secret key K. AES encryption involves multiple rounds of substitution, permutation, and mixing operations.

$$C = \text{AES\_Encrypt}(P, K)$$

Cryptography is foundational to secure communication, e-commerce, and data protection. The principles of number theory, as exemplified by algorithms like RSA, form the backbone of these cryptographic systems.

5.2   Number Theory in Computer Science

Number theory, the study of integers and their properties, has profound implications and applications in the field of computer science. In this section, we explore the intersection of number theory and computer science, showcasing how these mathematical concepts are crucial for algorithm design and optimization.

1. Cryptographic Hash Functions

Cryptographic hash functions are widely used in computer science for data integrity verification and password storage. A secure hash function, by design, is resistant to collisions and provides a unique output for each distinct input.

A fundamental property of hash functions is the avalanche effect, where a small change in the input results in a significantly different output. Many cryptographic hash functions rely on prime numbers and modular arithmetic to achieve this property.

$$H(x) = (a \cdot x + b) \mod p$$

Here, a, b, and p are carefully chosen prime numbers.

2. Primality Testing

Primality testing, the determination of whether a given number is prime, is a fundamental problem in number theory with direct applications in computer science. Algorithms like the Miller-Rabin primality test efficiently determine whether a number is composite or probably prime.

$$a^{n-1} \equiv 1 \pmod{n}$$

The test involves repeated modular exponentiations, and its accuracy can be controlled by the choice of the base a.

3. Random Number Generation

Random number generation is a critical aspect of many computer algorithms and simulations. The quality of randomness is often evaluated using properties related to number theory.

Linear congruential generators, a class of pseudo-random number generators, utilize modular arithmetic to generate sequences of seemingly random numbers.

$$X_{n+1} = (a \cdot X_n + b) \mod m$$

Choosing suitable values for a, b, and m ensures the statistical properties of the generated sequence.

4. Error Detection and Correction

Error detection and correction codes, essential in data storage and communication, leverage number theory concepts. For example, the Hamming code ensures the integrity of transmitted data by introducing redundancy based on powers of two.

$$2^r \geq m + r + 1$$

Here, r is the number of redundant bits, and m is the number of data bits.

5. Number-Theoretic Transformations

Number-theoretic transformations play a crucial role in computer science algorithms, especially in signal processing and cryptography. The Fast Fourier Transform (FFT), a fundamental algorithm for signal processing, relies on number-theoretic properties to efficiently compute the Discrete Fourier Transform.

$$X_k = \sum_{n=0}^{N-1} x_n \cdot e^{-i2\pi kn/N}$$

The roots of unity and modular arithmetic contribute to the efficiency of the FFT.

6. Euclidean Algorithm in Computer Science

The Euclidean Algorithm, a classical algorithm for finding the greatest common divisor (GCD), is widely used in computer science. It forms the basis for the Extended Euclidean Algorithm, crucial in modular inverses and solving linear Diophantine equations.

$$\text{GCD}(a, b) = \text{GCD}(b, a \mod b)$$

The Euclidean Algorithm is a fundamental tool for efficient algorithm design in computer science.

7. Number Theory in Algorithms Complexity Analysis

Number theory concepts are often employed in the analysis of algorithms' time complexity. The study of modular arithmetic and number-theoretic functions contributes to understanding the efficiency and scalability of algorithms.

For example, the time complexity analysis of certain algorithms involves studying the growth rate of functions, and number-theoretic functions like the totient function ($\phi(n)$) may appear in such analyses.

8. Public-Key Cryptography

Public-key cryptography, a cornerstone in securing online communication, relies heavily on number theory. The Diffie-Hellman key exchange algorithm, which allows two parties to establish a shared secret key over an insecure channel, utilizes modular exponentiation.

$$A \equiv g^a \pmod{p}$$

Here, g is a generator, a is a secret key, and p is a prime number.

9. Chinese Remainder Theorem (CRT) in Computer Science

The Chinese Remainder Theorem (CRT) finds applications in computer science, particularly in modular arithmetic and cryptography. CRT allows for efficient modular reduction and reconstruction in certain scenarios.

Given a system of modular congruences:

$$x \equiv a_1 \pmod{m_1}, \ x \equiv a_2 \pmod{m_2}, \ \ldots, \ x \equiv a_k \pmod{m_k}$$

CRT provides a solution x that satisfies all these congruences.

10. Quadratic Residues in Cryptography

Quadratic residues, a concept from number theory, find applications in cryptography, particularly in the design of cryptographic protocols.

The quadratic residue symbol $\left(\frac{a}{p}\right)$ determines whether a is a square modulo a prime p. This property is essential in protocols like the Quadratic Residue Diffie-Hellman (QR-DH) key exchange.

$$\left(\frac{a}{p}\right) = a^{\frac{p-1}{2}} \pmod{p}$$

These examples highlight the intimate connection between number theory and computer science. The elegance and versatility of number theoretic algorithms provide a rich toolkit for solving diverse problems in the realm of computing.

Appendix A

Number Theoretical Algorithms

In this appendix, we delve into additional mathematical concepts and examples that complement the main content of the book. These supplementary materials aim to provide a deeper understanding of number theoretic algorithms and their applications.

1. Continued Fractions

Continued fractions are expressions of the form $a_0 + \cfrac{1}{a_1 + \cfrac{1}{a_2 + \cfrac{1}{\ddots}}}$. They play a significant role in number theory, particularly in approximating real numbers. For example, the continued fraction representation of the square root of 2 is $[1; 2, 2, 2, \ldots]$.

2. Mersenne Primes

Mersenne primes are prime numbers that can be written in the form $2^n - 1$. A famous example is $2^{31} - 1$, which is the largest known Mersenne prime as of my knowledge cutoff in 2022.

3. Pell's Equation

Pell's equation, $x^2 - Ny^2 = 1$, where N is a nonsquare positive integer, has solutions in positive integers x and y. For example, for $N = 2$, the solutions are $(3, 2)$ and $(17, 12)$.

4. Elliptic Curve Cryptography

Elliptic Curve Cryptography (ECC) is a modern cryptographic technique based on the mathematics of elliptic curves. The security of ECC relies on the difficulty of the elliptic curve discrete logarithm problem. Elliptic curves have the form $y^2 = x^3 + ax + b$.

5. Riemann Hypothesis

The Riemann Hypothesis, one of the most famous unsolved problems in mathematics, is related to the distribution of prime numbers. It posits that all non-trivial zeros of the Riemann zeta function lie on the critical line in the complex plane.

6. Quadratic Forms

Quadratic forms, expressions of the form $ax^2 + bxy + cy^2$, are extensively studied in number theory. They have connections to the theory of quadratic residues and solutions to Diophantine equations.

7. Modular Forms

Modular forms are complex functions with certain transformation properties under modular transformations. They are important in advanced areas of number theory, including the proof of Fermat's Last Theorem.

8. Artin's Conjecture

Artin's Conjecture is a proposition in algebraic number theory, stating that for a fixed integer a that is not a perfect square, the quadratic polynomial $x^2 - a$ represents infinitely many prime numbers.

9. Class Field Theory

Class Field Theory is a branch of algebraic number theory that provides a deep understanding of abelian extensions of number fields. It relates to the distribution of prime ideals in these extensions.

10. Analytic Number Theory

Analytic Number Theory involves the application of methods from mathematical analysis to study properties of the integers. The Prime Number Theorem, which describes the asymptotic distribution of prime numbers, is a key result in this field.

11. Hensel's Lemma

Hensel's Lemma is a powerful tool for lifting solutions of polynomial congruences. It allows for the construction of solutions in modular arithmetic based on solutions modulo powers of prime numbers.

12. Lucas-Lehmer Test

The Lucas-Lehmer test is used to determine whether a Mersenne number $2^n - 1$ is prime. It exploits the properties of the Lucas sequence and plays a crucial role in the search for large prime numbers.

13. Legendre Symbol

The Legendre symbol $\left(\frac{a}{p}\right)$ is a number-theoretic function used in the study of quadratic residues. It reveals whether a is a quadratic residue modulo a prime p.

14. Randomized Algorithms in Number Theory

Randomized algorithms, such as the Miller-Rabin primality test, are employed in number theory for probabilistic primality testing. These algorithms provide fast and efficient ways to determine the primality of large numbers.

15. ABC Conjecture

The ABC Conjecture is a recent and intriguing hypothesis in number theory, linking the prime factors of three integers a, b, c such that $a + b = c$. The conjecture has deep connections to various branches of mathematics.

These topics, though not exhaustive, contribute to the richness of number theory. Exploring these mathematical concepts enhances the reader's understanding and appreciation for the breadth and depth of the subject.

Appendix B

Number Theoretical Algorithms

In this appendix, we explore additional topics in number theory, providing readers with supplementary insights and examples to further enhance their understanding.

1. Lucas Numbers

Lucas numbers are a sequence similar to the Fibonacci numbers, defined by the recurrence relation $L_n = L_{n-1} + L_{n-2}$ with initial values $L_0 = 2$ and $L_1 = 1$. The ratio of consecutive Lucas numbers converges to the golden ratio ϕ, just like in the Fibonacci sequence.

2. Thue-Morse Sequence

The Thue-Morse sequence is an infinite binary sequence with no overlapping blocks of consecutive digits being identical. It is constructed by repeatedly appending the complement of the existing sequence.

3. Carmichael Numbers

Carmichael numbers are composite integers that satisfy the modular arithmetic property $a^{n-1} \equiv 1 \pmod{n}$ for all a coprime to n, where n is the Carmichael number. They are a fascinating aspect of number theory related to Fermat's Little Theorem.

4. Farey Sequences

Farey sequences are sets of fractions between 0 and 1 with denominators less than or equal to a given bound. They are named after the British geologist John Farey, who introduced them in the early 19th century. Farey sequences have applications in Diophantine approximation.

5. Polyominoes and Pentominoes

Polyominoes are plane geometric figures formed by joining one or more equal squares edge to edge. Pentominoes are a specific type of polyomino consisting of five squares. Exploring their properties involves combinatorics and recreational mathematics.

6. Dedekind Zeta Function

The Dedekind zeta function is an extension of the Riemann zeta function, defined for certain number fields. It plays a crucial role in algebraic number theory, linking the properties of prime ideals to the distribution of prime numbers.

7. Voronoi Diagrams in Number Theory

Voronoi diagrams, derived from a set of points in a plane, are used in number theory to study the distribution of prime numbers. The positions of points in a Voronoi diagram are related to the distribution of primes in the number line.

8. Farey-Tarry-Harborth Constant

The Farey-Tarry-Harborth constant is a mathematical constant representing the limiting density of fractions not exceeding 1 in reduced Farey sequences. It has connections to Diophantine approximation and the theory of continued fractions.

9. Gauss's Circle Problem

Gauss's Circle Problem is a classic problem in number theory, involving the estimation of the number of lattice points within a large circle. It showcases the interplay between number theory and geometry.

10. Brocard's Problem

Brocard's problem involves finding integer solutions to the Diophantine equation $n! + 1 = m^2$. It is named after Henri Brocard, who investigated it in the late 19th century. The solutions to this problem have connections to factorials and perfect squares.

11. The Collatz Conjecture

The Collatz Conjecture, also known as the 3n+1 problem, involves iterating a simple mathematical operation on integers. Despite its simplicity, the conjecture remains unsolved, making it one of the most famous open problems in mathematics.

12. Arithmetic Geometry

Arithmetic geometry explores the connections between algebraic geometry and number theory. It involves studying algebraic varieties over number fields and their rational points. The celebrated Mordell conjecture is a prominent topic in arithmetic geometry.

13. Perfect Numbers

Perfect numbers are positive integers that are equal to the sum of their proper divisors, excluding the number itself. The first few perfect numbers are 6, 28, and 496. Euclid proved that every even perfect number has a specific form related to Mersenne primes.

14. Stirling Numbers

Stirling numbers of the first and second kind arise in combinatorics and calculus. They count the number of ways to partition a set and the number of permutations with a fixed number of cycles, respectively.

15. The Twin Prime Conjecture

The Twin Prime Conjecture posits that there are infinitely many twin primes, pairs of primes that differ by 2. Despite extensive computational evidence supporting the conjecture, a proof remains elusive.

These topics provide additional depth and breadth to the study of number theory. Exploring these mathematical concepts offers readers a broader perspective on the richness of the subject.

Bibliography

In this bibliography, we list key references that have greatly influenced the content of this book on Number Theoretic Algorithms. The field of number theory is vast, and these works provide a foundation for understanding the various algorithms and concepts discussed.

1. Hardy, G. H., & Wright, E. M. (2008). An Introduction to the Theory of Numbers. This classic work serves as a comprehensive introduction to number theory, covering topics such as Diophantine equations, prime number theorems, and modular forms.

2. Ireland, K., & Rosen, M. (2013). A Classical Introduction to Modern Number Theory. The book explores classical topics in number theory, including Fermat's Last Theorem, quadratic forms, and the distribution of prime numbers.

3. Knuth, D. E. (1997). The Art of Computer Programming, Volume 2: Seminumerical Algorithms. This volume delves into algorithms, including those related to number theory. Knuth provides a thorough exploration of various numerical methods.

4. Silverman, J. H. (2009). A Friendly Introduction to Number Theory. The book presents an accessible introduction to number theory, covering topics like modular forms, elliptic curves, and the ABC conjecture.

5. Apostol, T. M. (1976). Introduction to Analytic Number Theory. This text focuses on analytic methods in number theory, including the distribution of prime numbers and the Riemann zeta function.

6. Shoup, V. (2005). A Computational Introduction to Number Theory and Algebra. The book provides a computational perspective on number theory, introducing algorithms and methods useful for practical applications.

7. Ribenboim, P. (2000). The Book of Prime Number Records. This compilation gathers records and interesting facts about prime numbers, showcasing the diversity and complexity of prime number patterns.

8. Mollin, R. A. (2008). RSA and Public-Key Cryptography. Mollin's work explores the mathematical foundations of RSA cryptography, providing insights into the underlying number theory.

9. Hardy, G. H., & Littlewood, J. E. (1923). Some problems of Diophantine approximation: Part I. This landmark paper initiated the study of Diophantine approximation, a branch of number theory dealing with the approximation of real numbers by rational numbers.

10. Serre, J. P. (1973). A Course in Arithmetic. Serre's influential book covers various aspects of algebraic number theory, including rings of integers, class field theory, and the zeta function.

11. Lenstra, H. W., & Lenstra, A. K. (1992). Algorithms in Number Theory. The authors provide an overview of algorithms in number theory, covering topics such as factoring, primality testing, and elliptic curve cryptography.

12. Ribenboim, P. (1996). The Little Book of Big Primes. This book is a delightful exploration of large prime numbers, their properties, and their significance in number theory.

13. Davenport, H. (2000). Multiplicative Number Theory. Davenport's work is a classic in multiplicative number theory, covering topics like the distribution of prime numbers in arithmetic progressions.

14. Cohen, H. (2007). Number Theory: Volume I: Tools and Diophantine Equations. Cohen's comprehensive treatment of number theory introduces tools and techniques for solving Diophantine equations.

15. Lenstra, A. K., Lenstra, H. W., & Lovász, L. (1982). Factoring polynomials with rational coefficients. This seminal paper introduces the Lenstra–Lenstra–Lovász algorithm for factoring polynomials over the rational numbers.